ASE Lab Books

ENERGY
and CHEMISTRY

Compiled by
W. J. HUGHES

John Murray   Albemarle Street   London

Printed in Great Britain by Butler & Tanner Ltd,
Frome and London

0 7195 3094 6

ASE Lab Books

ENERGY
and CHEMISTRY

ASE Lab BOOKS
General editor   A. A. Bishop

Biology editors   D. G. Mackean and A. Davies
Chemistry editors   Miss F. Eastwood and E. H. Coulson
Physics editors   D. Shires and M. F. James
Middle Schools editor   Eric Deeson

Other titles in this series

BIOLOGY
Plant Physiology compiled by C. J. Clegg
Ecology compiled by A. Davies
Cytology, Genetics and Evolution compiled by G. W. Shaw

CHEMISTRY
Chemical Equilibrium, Acids and Bases compiled by J. M. Newman
Chromatography compiled by R. Worley
Catalysis compiled by D. G. Newman

MIDDLE SCHOOL
Biology and Chemistry for Middle Schools compiled by J. Bushell
Physics and the Earth Sciences for Middle Schools compiled by E. Deeson

PHYSICS
Mechanics and Properties of Matter compiled by J. C. Siddons
Light compiled by E. Deeson
Heat compiled by W. K. Mace
Electronics compiled by E. W. Mackman

Contents

Introduction

The interaction between energy and matter is obvious from the earliest stages of chemical investigation and many experiments can be used, incidentally, as examples of processes in which energy is transferred to the environment and vice versa. Discussion of energy changes accompanying material changes can begin from the moment of lighting the Bunsen burner for the first time. The combination of the less reactive metals with sulphur (e.g. Fe, Cu), the reaction of acids with alkali, and many other systems are equally useful.

Many thermochemical ideas can appropriately be introduced into chemistry courses. These include the following:

1. The different forms in which energy can be transferred between the system and the environment—heat, light, electricity, sound and mechanical energy. If we follow the changes far enough all forms of energy seem to end up as thermal energy. Chemists are sometimes interested in the energy transfer only as in the internal combustion engine. At other times the new products are of prime importance—the energy transfers are then of economic interest.
2. Some changes need to be started or 'triggered off' by a small amount of energy but then proceed on their own. Thus, lighting a burner or burning magnesium might lead us to a consideration of activation energy and possibly catalysis. Ignition temperature experiments are of the same kind.
3. Other changes occur without external intervention. These are so-called spontaneous changes and include such phenomena as dissolution, neutralization, spontaneous combustion, and rusting.
4. Yet other changes proceed only if energy, in some form, is provided continuously by the environment. Non-spontaneous changes include electrolysis experiments and many thermal decomposition reactions.

Energy must be supplied by the environment in order to boil, melt, evaporate or distil. The reversibility of such processes is brought out when the energy is 'paid back', when a vapour condenses or a liquid freezes, for example.

The fact that heat is supplied to make anhydrous copper(II) sulphate(VI) from the hydrated crystals and that a heat bonus appears when the products are mixed is an excellent example of the interchange between the system and the environment. The economics of the same reaction is worthy of early consideration.

5. Change of state is largely concerned with the separation of particles; the breaking of bonds is not involved, though exceptions include the rupture of hydrogen bonds as in changing water into steam.

   Atoms form molecules or giant structures and energy is transferred to break up these combinations. Thus bond-breaking is involved in the heating of sulphur and burning of carbon in contrast to the heating of iodine crystals. The contrast between burning and melting of naphthalene, for example, is well worth consideration.

6. Energy changes whether exothermic or endothermic are often well summarized by means of energy-level diagrams.

7. That energy changes are additive is a fact taken for granted. It is fundamental in all calorimetry and any quantitative work involving Hess's law and Born–Haber cycles.

8. There is a difference in the amount of heat and the amount of work that a reaction can furnish, and consideration of this inequality is one possible way of leading to the entropy concept.

The experiments that follow by no means cover all the aspects of thermal energy enumerated above and there is an implied invitation to chemistry teachers to make the second edition of this booklet more comprehensive and therefore more valuable.

W. J. H.

Energy transfer

Heat, the movement of molecules
AUDREY H. HEAP

The experiment follows an investigation of the diffusion of gases and of diffusion in liquids and gels. From the results it is argued that in liquids and gases molecules are in perpetual motion. Almost fill two similar beakers, A and B, with water and support them side by side, A on a retort stand for heating, and B on a wooden block. Place a second block alongside A. Clamp a tube upright in the middle of B with its end about 0.3 cm off the bottom, so that a crystal dropped into the tube rests in the centre. Drop in a crystal of potassium manganate(VII). Obviously it is denser than water, yet, in a few moments, the manganate(VII) can be seen rising in the beaker, and presently is evenly distributed throughout the liquid. This can have come about only by collisions with the moving water molecules. When the water in A has come to the boil, rapidly transfer the beaker to the wooden block—to minimize convection currents—and drop in a crystal of manganate(VII). Instantly, or in a few seconds, according to the size of the beaker, the pink colour is seen evenly distributed throughout the liquid. The molecules of hot water are therefore moving much faster than those of cold water. From these experiments it is easy to give a physical reality to heat and temperature besides obtaining a connection between chemistry and physics.

Energy conversion using a drawing pin
E. J. HARRIS

A brass drawing pin is pushed well in to the end of a cork and rubbed vigorously to and fro on a wooden surface for about ten seconds. The energy of the operator is converted into thermal energy in the drawing pin and its temperature rises. The rise of temperature can be demonstrated by placing the drawing pin, immediately after rubbing, on the junction of a thermocouple connected to a sensitive galvanometer. It is advisable to make the thermocouple of *thin* wire, e.g. 32 gauge copper and 32 gauge constantan, so that the thermal capacity of the junction can be kept small.

To get good results the drawing pin should be pressed well down on to the wood while it is being rubbed to and fro, and fairly long strokes should be used—about 15 or 20 cm. The temperature rise is quite considerable, because the thermal capacity of the drawing pin is extremely small and the cork is a fairly good thermal insulator. It is convenient to place the junction on one or two sheets of blotting paper on the bench and then press the hot drawing pin well down on to the junction. An e.m.f. of the order of 1 or 2 millivolts can be produced fairly easily.

If desired the experiment could, alternatively or additionally, be done by the whole class in pairs, using the hand as a temperature detector instead of a thermocouple. If the drawing pin is placed on the back of the hand the temperature rise is very noticeable; in fact, a 10-seconds rubbing could easily produce an unnecessary amount of pain and it might be advisable to use a shorter time to begin with. It is also possible to produce a blue coloration on a heat-sensitive paper, but I find this involves considerably more work on the part of the operator!

Light energy controlling sound energy

J. C. SIDDONS

A neon bulb is in parallel with a capacitor, whose capacitance is of the order of 0.02 microfarad, and earphones. This arrangement in turn is in series with a photo-electric cell, e.g., a selenium cell. Every time the capacitor potential

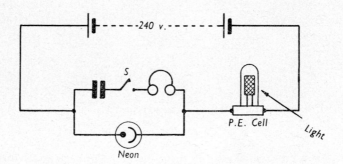

reaches the sparking potential of the neon a flash occurs and a click is heard (though 'cluck' more accurately describes the sound). When no light falls on the cell few clucks are heard, but as the light is increased more and more frequent become the clucks. Selenium cells have a finite resistance in the dark, but caesium cells have almost infinite resistance then. With caesium cells the clucks should be very rare when no light falls.

The clucking rate can be altered also by altering the capacitor. If the switch S is opened and light is falling on the selenium a high pitched note is heard. This illustrates the ability of alternating currents to surge into and out of electrical culs-de-sac.

The experiment is not original, I came across it in the Russian-language journal, *Physics in School*. I use it with junior forms to show how light energy via electricity can produce sound energy but it can be used also as a photometer, for if two sources of light at the same distance from the cell give equal clucking rates then they have equal candle powers.

In place of the selenium cell a light-dependent resistor (such as the cadmium sulphide ones supplied with electronic kits) can be used: if in place of the earphones an amplifier is used much louder clucks can be made.

A simple heat engine
J. C. COOPER

The apparatus demonstrates the conversion of heat energy to mechanical energy.

On heating the bi-metallic strip, ABC, with a lighted match, say, its curvature decreases so that end A advances up the slope, C remaining fixed in

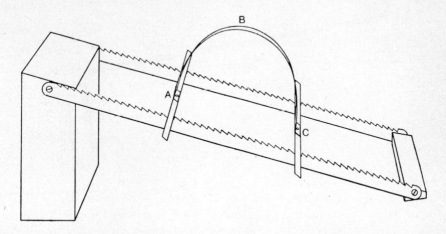

position owing to the shape of the teeth. On allowing to cool, end A remains fixed and C advances. The apparatus climbs about 2.5 cm up the slope, work being done against gravity. The teeth of the hacksaw blades are shown much enlarged.

The Telegraph Construction and Maintenance Co. kindly supplied a specimen of their Bi-metal 140 of thickness 0.05 cm and width 0.6 cm. Two

0.6 cm wide 12.5 cm lengths of tin plate cut from a tin can were riveted to a 30 cm length of bi-metal to form a letter H. The strip was then bent to form a semicircle with the metal which expands most on the inside of the curve. Further sharp bends of the strip at A and C, as shown, make the movement rather smoother. 30-cm hacksaw blades of 6.5 teeth per cm are satisfactory for the ramp. Grinding off the sharp edges of the teeth reduces jerkiness of movement as does removal of sharp edges from the tin-plate strips.

The bi-metal must not be heated above 500 °C.

A pure heat engine

W. K. MACE

In the introductory teaching of 'forms of energy' the conversion of heat into work is most commonly illustrated with reference to the steam engine. As a clear-cut example of the conversion in question this engine leaves much to be

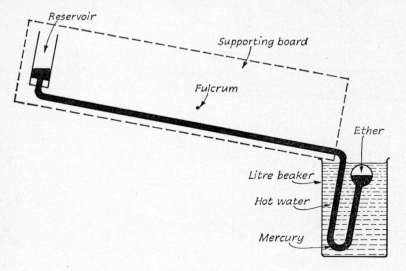

desired. I know that when I was a child, whatever the textbooks might say, a steam engine to me was a thing which took in water and coal and gave out work, steam, smoke and ash; all outputs and inputs were equally prominent. In fact, I doubt if I really understood heat engines properly until many years later, when I acquired a hot-air engine to play with. Suddenly it was clear: you took it out of the cupboard, played a Bunsen on it, and work came out!

My own view is that all 'energy conversion kits' should include a hot-air engine. Unfortunately there does not seem to be a suitable cheap toy one on the market, so one has to make one. If this should be impossible, the ether engine illustrated above is in many ways a fair substitute. All you need to do is take it out of the cupboard and give it a beaker of hot water. It immediately starts doing work.

Heat is absorbed by the ether which evaporates and pushes mercury over into the reservoir. The tube tips, the bulb comes out into the air and cools; the mercury runs back, re-immersing the bulb in the water. The cycle repeats indefinitely.

The tube shape is dictated by the problem of getting the optimum amount of ether. Initially one puts in too much, and excess then finds its way out to the reservoir surface. Stability is a nice problem: in practice the position for the fulcrum has to be found by trial and error.

A simple demonstration steam turbine

H. LANGDON-DAVIES

ABC is a bent copper tube whose outside diameter is about 0.6 cm. The portion AB is well soldered to a base plate D, and the plate is then screwed to a board large enough to keep the apparatus steady.

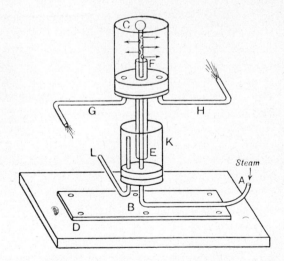

Near the top of the tube a number of steam inlet holes are drilled, and at the very end a steel ball-bearing is placed. The ball-bearing takes the weight of the rotor, and the tube BC acts as a spindle.

To construct the rotor a small can (or calorimeter, or glass tube) is fitted with a cork. The cork is drilled centrally to take the glass tube EF, and two other holes are made to take the tubes G and H. The tube EF should fit the copper tube closely, and yet should be free to rotate. It may be necessary to try several tubes before the right one is found. The tubes G and H are about 13 cm long, and are bent twice at right-angles, as shown. The extremities of these tubes are drawn out slightly to form the orifices for the steam jets. If care is exercised, it is possible to balance the rotor sufficiently well to allow high speeds to be attained when steam is passed up the tube BC.

Some steam will escape through the sleeve EF, and more power will be attained if the end of the sleeve is made to dip under mercury. The method of doing this is obvious from the diagram, the tube K being three-quarters full of mercury, L is a draining-tube which takes away the surplus water which tends to collect on the top of the mercury.

The rotor can be withdrawn from the spindle in order to explain the working of the model, and at the close of the demonstration the ball-bearing can be dried and stored for further use.

Hero's engine
G. W. YOUNG

A simple and inexpensive form of this 2000-year-old toy—the earliest known steam engine—can be made quite easily from scrap material, using only the simplest tools (soldering iron, files, tin-snips and pliers).

Required: a vacuum-type tobacco tin, B, about 7 cm diameter, to form the boiler, which is also the rotating member; a somewhat wider tin, A, for the base, which is also the spirit-lamp; about 15 cm of copper tubing a little over 1 mm bore and with rather thick walls to facilitate bending without kinking (i.e., some 3 mm external diameter) and a small quantity of sheet tin, cut from old tin containers and hammered flat. E is a shortened steel knitting needle (9 cm) to form the spindle on which B rotates.

First solder on the lid of B. Then find its centre as accurately as possible and drive the point of the knitting needle through to touch the bottom of the tin. Making sure that the needle is held perpendicular to the surface of the lid, drive it through the bottom of the tin until it projects about 2.5 cm. If this has been done carefully, the tin should run true on twirling the spindle; any 'wobble' will detract greatly from the appearance of the finished model when running, so care here is essential. The spindle can now be secured by soldering. The jet-tubes, D, are each made from about 4 cm of the copper tubing and are bent to a gentle curve so that the steam-jets will issue as nearly as possible tangentially to the edge of B. The inner ends of these tubes are inserted fairly near the centre of the lid—if fixed near the circumference, water would of course be driven out by centrifugal force later. The holes for the tubes are made with the point of a wire nail used as a punch. If the ends of the tubes are filed slightly conical, they can be made a tight fit in these holes and thus held securely for soldering. The filling-tube, F, is made by wrapping a narrow strip of tin round a pencil and soldering the seam. Solder this tube to the lid, punch the hole with the nail, and enlarge this hole (using the tang of a file as a reamer) until it is roughly the internal diameter of the tube. The latter is closed in use by *lightly* inserting a small cork, and this constitutes a simple safety device. It is well at this stage to blow through F in order to make sure that the jet-tubes D are free from obstruction.

The lid of the spirit-container, A, is not soldered on, but should be a reasonably good fit. The short tube, H, leads any accumulation of vapour

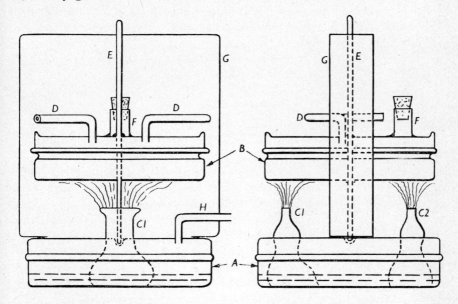

away from the flame. The two burners, C1 and C2, are made and fitted as described for the filling-tube, F. They are fixed well away from the centre of the lamp so that the flames may play on the *outer* part only of the base of the boiler, and avoid contact with the soldered joint fixing the spindle. This precaution is necessary since, when running, B soon attains a speed of 250–300 r.p.m., at which speed the centre of the base is probably quite dry, and direct contact with the flames would melt the solder. After fixing, the top of each burner is squeezed with the pliers to form a narrow slot, and this gives a longer and narrower form to the flames, thus helping to keep them away from the centre as required. Cotton-wool, packed firmly into each tube, forms a suitable wick. Large flames are not needed. Steam-raising with about 40 cm³ of water in B takes 2–3 minutes and the model runs for 5–7 minutes on one filling.

The lower bearing of the spindle is merely a very shallow depression made in the lid of A—again using the wire nail, but giving only a *gentle* tap with the hammer. This simple bearing gives a remarkably free-running spindle, and on this much of the success of the model depends. The upper bearing is a hole punched through the strip of tin, G, which is bent to form a bridge whose lower ends are soldered to the lid of A. The height of G should be such that due clearance is allowed for the removal of B when necessary—say 8 cm.

Painting the model in two colours with a modern steam-resisting enamel adds a finishing touch.

Notes: Method and tools have been kept as simple as possible so that the job might be tackled by any boy who could use a soldering iron. The boiler should be emptied after use in order to check rusting; in any case the fittings are easily transferred to a new tin should this ultimately become necessary. The writer's model has done much running over a considerable period, with no trouble and no renewals.

Lemery's volcano

IOLO WYN WILLIAMS and K. R. RAWLINGS

The reaction of sulphur and iron filings must be one of the most frequently demonstrated, but we wonder how many teachers know that sulphur and iron will give iron(ii) sulphide if the mixture is simply rubbed into a paste with water. If this seems incredible it is quite easily verified. We made up a mixture of 21 g of iron powder and 12 g flowers of sulphur, then stirred in 6 cm³ water. The temperature of the mixture rose slowly, then more quickly, then very rapidly indeed to 100 °C, the last 20 degrees taking no more than a few seconds; the whole process took about six minutes. Steam belched forth accompanied by a slight sulphurous smell, and after three more minutes, by which time the product had become a fine, dry, black powder, the temperature rose slowly again to 105 °C, where it remained for three or four minutes before cooling to room temperature.

This experiment was first performed by Nicholas Lemery (1645–1715), a notable French chemist who was particularly interested in geological phenomena. It is said that he buried in the ground a moist mixture of fifty pounds of sulphur and iron filings. After some hours the mixture burst forth, as an artificial volcano, complete with fire and fume. In this form, or on a smaller scale, the experiment features in most practical chemistry books up to the end of the nineteenth century but now appears to have been completely forgotten.

The product is a fine black powder, which is strongly magnetic. This led us to look at it rather more closely, and to carry out a semi-quantitative experiment to see whether it is in fact iron(ii) sulphide. Apart from a slight sulphurous smell during the reaction there was no evolution of hydrogen sulphide or of sulphur dioxide. Most of the water appeared to be driven off as steam, and with the proportions already given the product weighed 1.4 g more than the reactant iron and sulphur. The product gives a copious supply of hydrogen sulphide when reacted with hydrochloric acid; 0.32 g yielded 68 cm³ gas with 5 M HCl, leaving about 0.06 g of a black residue. At least 90 per cent of the gas was absorbed in lead acetate solution. These results are fully compatible with the evolution of hydrogen sulphide from iron(II) sulphide, though the magnetism is anomalous.

Gas, vapour and dust explosions

R. G. BRAY

Impressive demonstrations of gas, vapour and dust explosions are possible using polythene bags filled with the fuel/oxygen mixtures. The explosions are free from dangerous flying solids, but do satisfy the destructive urge!

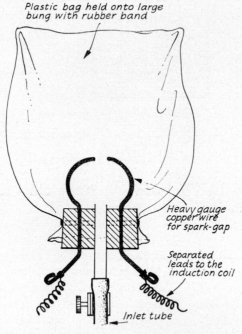

Fig. 1. General assembly

For indoor experiments a medium-size plastic sandwich-bag is filled with oxygen and the fuel, supported in a stand, and then fired in a darkened room with a spark from a distant induction-coil (Fig. 1). The most successful fuels tried so far have been coal gas, hydrogen, ammonia, petrol vapour, meths vapour, and finely powdered charcoal. A wide range of fuel/oxygen volume ratios can be tried using a gas-pipette and the liquid fuels may be conveniently vaporized into the oxygen using the apparatus shown in Fig. 2. Excess pulverized solid fuels should be placed in the deflated bag before admission of the oxygen and should be shaken up just before firing.

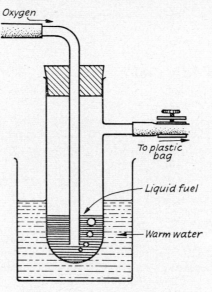

Fig. 2. Vaporizer for liquid fuels

Polythene flask used in full view of the class for the oxygen-hydrogen explosion

J. W. DAVIS

The flask, 16.5 cm tall, 8 cm wide, neck-width 3 cm, fitted with a cap and cup, obtainable for domestic use, is filled with water and held inverted in water. Oxygen is bubbled in to the level of the top of adhesive tape which surrounds the flask and indicates when the flask is one-third full of the gas, and hydrogen is bubbled in to fill the rest of the flask. The cap is fitted and the flask put upright on the bench. The cap is removed and the gas lit. Of course, the flash, in addition to the bang, is perceived by the onlookers.

The explosion of gaseous mixtures

P. D. ARCULUS

After several times seeing and hearing Colonel B. D. Shaw of Nottingham University giving his superb experimental lecture on 'Explosives', I have employed his safe method for exploding gaseous mixtures.

The dimensions, which are not critical, and the arrangement of the apparatus are illustrated in the accompanying diagram.

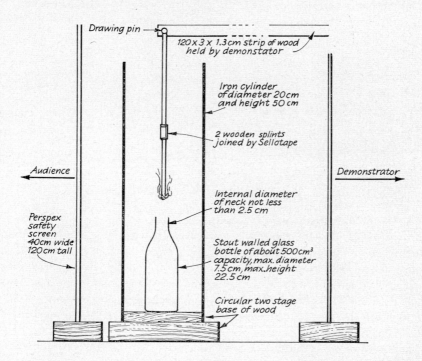

A bottle, calibrated for a particular explosion mixture either by painting a ring round it [1] or, more rapidly but not permanently, by using a 'Flowmaster' felt nib pen for the same purpose, is filled with the gases of the explosion mixture and the neck closed with a rubber bung. The equipment is assembled in a stair well or alternative open space [1], the bung is removed, and a well inflamed wooden splint thrust centrally into the iron cylinder. The explosion may not be immediate but will be safe and resounding. With an acetylene–oxygen mixture a new bottle will be required if the experiment is to be repeated!

REFERENCE

1. Holt, Charles, *The School Science Review*, 1962, **44**, 161.

On combining heat content diagrams

R. J. SWAN

(1) Certain principles of thermochemistry appear to have been understood in this part of the world for a good 1700 years. The Babylonian Rabbi Judah advised people who went to the hot baths to 'open your mouths and expel the heat', which fitted in with Rabbi Samuel's dictum 'heat expels heat' [1]. Eight hundred years later the French scholar, Rabbi Solomon ben Isaac, explained that 'when you open your mouths the heat of the baths goes into the body, and expels the heat of perspiration'. We could demonstrate this advice in school with a composite heat content diagram:

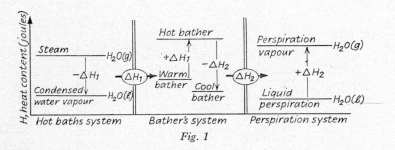

Fig. 1

This type of diagram not only emphasizes the change in heat content of each system, but also the *transfer of heat* from one system to another. Similar diagrams do not appear to be used in the common textbooks, although diagrams showing heat transfer from one system to another, but *without* simultaneous discussion of heat content changes, are used [2, 3]. The aim of this article is to show how much these composite diagrams can help pupils understand certain thermochemical problems. At a later stage they should help in discussions on entropy changes.

The essential point is that we always discuss two systems simultaneously, so that it is always clear that heat is released by one system and absorbed by the other. In the diagram we can see that an *exo*thermic change really involves the *exit* of heat.

(2) For these diagrams to be of any use, the pupil must understand what is meant by 'heat content'. This is discussed in some detail in the Nuffield *Handbook* [2]. Here it is sufficient to reiterate the point that *any* change in heat content of *any* non-reacting system (e.g. cooling of test-tube glass, heating up of air) involves a molecular change just as much as in chemically reacting systems. Indeed the names 'system' and 'surroundings' are arbitrary.

(3) To introduce the pupil to these diagrams, we can take familiar reactions, such as food being cooked on the stove (Fig. 2). Practically anyone, even without formal chemical education, can grasp the principles involved. He

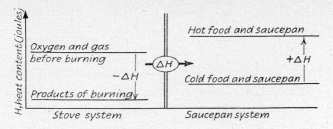

Fig. 2

can also grasp a more complicated diagram, as in Fig. 3, in which the loss of efficiency of the process is shown.

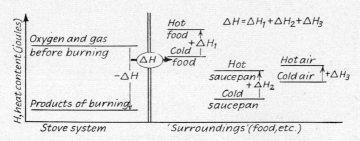

Fig. 3

(4) In the first demonstration (or laboratory) experiments in thermochemistry, the transfer of heat from one system to another must be emphasized. A suitable example is the solution of solid sodium hydroxide in water, in a test-tube which has iodine crystals stuck to its outside [4] (Fig. 4).

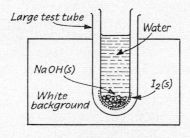

Fig. 4

The reaction is very exothermic, and the iodine vapour which collects in the large test-tube is easily seen. A simple diagram for the reaction would be as Fig. 5.

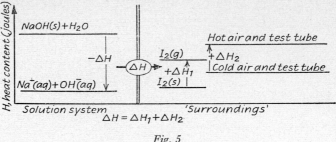

Fig. 5

To demonstrate endothermic changes, the reaction between solid barium hydroxide and solid ammonium thiocyanate is ideal. The vessel is placed on a drop of water on a piece of wood [5]. The water freezes and the vessel sticks to the wood; the heat content changes are summarized in Fig. 6.

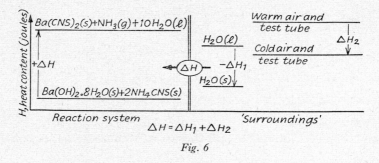

Fig. 6

In both these demonstrations it is clear that the exothermic change in one system is accompanied by an endothermic one in the other. Once this principle is firmly established, we should no longer hear the plaint, 'But how can it be an endothermic reaction, absorbing heat, when I can feel it getting cold?' (or vice versa). A composite diagram shows that we feel the temperature change (which accompanies the heat content change) of the surroundings, *not* the change in the heat content of the reacting system. For all that, *after completion of the reaction*, the two systems come to thermal equilibrium, and so it appears that—to take the example above—the reacting system has cooled.

A similar example is the query, 'Why, when I run and expend energy, do I get hot? Doesn't getting hot mean that I am absorbing energy? Indeed it does, but the distinction must be made between the reacting system (the muscles) and the surroundings (the rest of the body).

(5) This approach allows us to explain reactions in solution, wherein the reactants are the solute and the surroundings are the solvent. Before going into the problem, we should first analyse the composition of the reacting solutions.

For example, if it is the reaction of 1000 cm³ of NaOH 1.0 M with 1000 cm³ of HCl 1.0 M, the aqueous solutions consist of

$$1000 \text{ cm}^3 \text{ NaOH } 1.0 \text{ M}: \begin{cases} 1 \text{ mole NaOH, volume} \approx 20 \quad \text{cm}^3* \\ 55 \text{ moles } H_2O, \quad \text{volume} \approx 980 \text{ cm}^3 \end{cases}$$

$$1000 \text{ cm}^3 \text{ HCl } 1.0 \text{ M}: \quad \begin{cases} 1 \text{ mole HCl}, \quad \text{volume} \approx 20 \quad \text{cm}^3 \\ 55 \text{ moles } H_2O, \quad \text{volume} \approx 980 \text{ cm}^3 \end{cases}$$

Granted that part of the water is water of solvation, but the greatest part (90 per cent) certainly does not participate in the reaction

$$H^+(aq) + OH^-(aq) \rightarrow H_2O(l)$$

in which the water formed is no more than 1 per cent of the total quantity of water present. The thermometer in the solution measures the temperature rise of the non-reacting surroundings (the water)—Fig. 7.

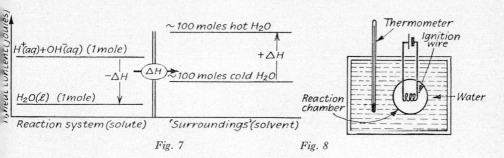

Fig. 7   Fig. 8

There is no difference in this approach from that used in measuring heat changes with a calorimeter—Fig. 8.

The reaction which takes place in the chamber, e.g. burning of fuel, releases heat to the surroundings, the water, and we measure the rise in temperature of this water. There is no intrinsic difference when using a 'homogeneous' calorimeter, in which there is no separation between the 'reaction chamber' and the surroundings. (We should note that the surroundings are no longer pure water, but a dilute solution of NaCl; we can neglect the possible change in thermal capacity of this solution relative to that of water.)

* The molar volumes of solids are not necessarily the same as those occupied when they are in solution. For this reason they are approximated here.

REFERENCES

1. *Babylonian Talmud*, Shabat, 41a (English translation: Soncino Press, London (1938), 1961).
2. Nuffield Chemistry, *Handbook for Teachers* (Longmans/Penguin Books, 1968), Ch. 16.
3. Rogers, M., *Energy in Chemistry* (Heinemann Educational Books, 1968).

4. 'Tested Demonstrations', *J. Chem. Educ.*, 1970, **47,** A206.
5. Alyea, H. N., and Dutton F. B. (editors), *Tested Demonstrations in Chemistry* (Chem. Educ. Publishing Co., Easton, Penn., 1965), p. 17.

Model to illustrate energy changes in chemical reactions

D. W. H. TRIPP

No originality is claimed for this model—a similar model was used in Professor Porter's recent television series, but details of construction and lettering may be of interest.

Two identical profiles are cut from plywood and are separated from each

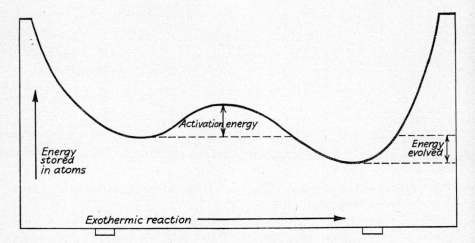

Fig. 1. Diagram of model to illustrate energy changes in chemical reactions. Exothermic side. The diagram is drawn to scale

other by two thicknesses of the same plywood which are placed in small pieces at 'strategic' places between the profiles.

60 cm of 2.5 cm × 0.95 cm wood in 15 cm lengths are fitted to small blocks set between the profiles so that they may be turned parallel to the profiles for storage. The model is painted with light gloss paint.

The model is reversible, one side being lettered for an exothermic change (Fig. 1) and the other for an endothermic change (Fig. 2). Lettering and lining were done with felt-markers and stencil pens, using different sizes of pen-stencils.

A large marble, 2.5 cm diameter, is placed on the rails formed by the profiles in the left-hand hollow which corresponds to the energy of the atoms of the reactants. It may be shown that in order to 'get over the hump' a certain minimum energy—the activation energy—is needed. If the profiles are suitably shaped (see details below), the model works both ways so that the reversibility of a chemical change may be appreciated.

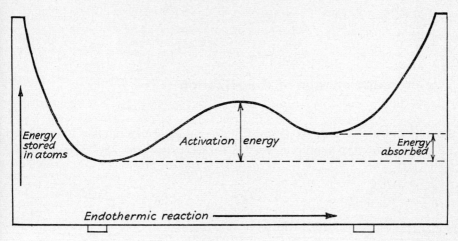

Fig. 2. *Diagram of model to illustrate energy changes in chemical reactions. Endothermic side. The diagram is drawn to scale*

Overall length—60 cm
Height at ends—28 cm
Height of higher hollow—12.5 cm
Height of lower hollow—8.5 cm
Height to hump—17 cm
Horizontal distance from left-hand edge to higher hollow (exothermio reaction)—15.7 cm
Horizontal distance from left-hand edge to hump—27.4 cm
Horizontal distance from left-hand edge to lower hollow—45 cm.

Measuring energy changes

The enthalpy change of vaporization

T. SAVORY

An experiment which demonstrates the fact that after water has been heated to 100 °C a further supply of heat is necessary to produce boiling is illustrated herewith.

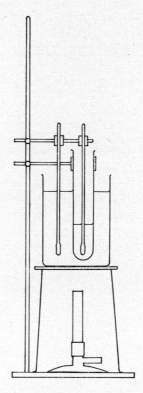

The water in the beaker boils and is seen to be at 100 °C. The water in the test-tube, heated by this boiling water, is seen to be also at 100 °C, but it does not boil. (Heat can pass from the flame at about 800 °C to the boiling water at 100 °C, but not from beaker to test-tube, because they are both at the same temperature.)

The experiment can be varied by boiling the water in the test-tube over a second Bunsen, and seeing that boiling stops when it is placed in the boiling water in the beaker.

Cheap calorimeter cases from waste polystyrene

P. T. HAYDOCK

We used polystyrene sheets that came as packing with apparatus such as balances. Rings were cut from the polystyrene with a hot wire. The rings were then built up to the required size and glued together.

The cases are found to be light, easy to use, and make very good insulators.

Lids to fit can be made in a similar way with holes for thermometer, stirrer and heating coil if required.

A simple experimental test of Hess's law

P. G. MATTHEWS

The object of this experiment is to find the heat of reaction between a metal and water, the heat of neutralization of the hydroxide produced and the heat of reaction between the metal and acid. According to Hess's Law, the last should be equal to the sum of the first two.

i.e.
$$Li + H_2O \rightarrow LiOH + \tfrac{1}{2}H_2; \Delta H_1$$
$$LiOH + HCl \rightarrow LiCl + H_2O; \Delta H_2$$
$$Li + HCl \rightarrow LiCl + \tfrac{1}{2}H_2; \Delta H_3$$
$$\Delta H_1 + \Delta H_2 = \Delta H_3$$

Suitable metals were found in lithium and calcium though slightly different techniques were necessary with the two. An ordinary copper calorimeter, capacity about 150 cm^3, was taken, weighed to the nearest gram and placed in a well lagged container. Into it were put 50 cm^3 of water with a pipette. A piece of lithium was cut off, a cube of about 3–4 mm was found suitable, and wrapped loosely in well perforated lead foil. This weighs down the metal, giving better mixing and less spitting. The temperature of the water was taken to 0.1 °C, the lithium was added and the maximum temperature of the well stirred mixture at the end of the reaction was found. Since the reaction is complete in about 30 s no cooling correction appeared necessary. When using calcium, a calcium turning was broken up in a mortar and about 0.3 g weighed roughly. This was added to the water, no lead foil being necessary, and as soon as the reaction began, a clock was started. The mixture was stirred vigorously throughout to prevent clogging by calcium hydroxide.

The reaction was timed and a rough cooling correction made by finding the temperature drop per minute at the maximum temperature, halving this and multiplying by the time of the reaction. The result was added to the maximum temperature.

To find the heat of neutralization, the calorimeter and contents were adjusted to the same temperature as some accurately standardized hydrochloric acid of molarity about 1.5. This can easily be done by cooling on ice to a temperature just below the acid and then warming with the hand. 25 cm³ of the acid were then added with a pipette and the rise in temperature noted. It is not strictly necessary to adjust the temperatures as described but it simplifies the calculation.

An allowance was made for the heat of dilution of the acid by finding the temperature rise, about 0.3° under these conditions, produced when 25 cm³ of acid were added to 50 cm³ of water at the same temperature.

To find the quantity of metal used, the contents of the calorimeter were made up to 250 cm³ and titrated against 0.1 M alkali. From the results the number of moles of metal was calculated.

The heat of each reaction was calculated by

$$\Delta H = \frac{\text{C. of calorimeter and contents} \times \text{Rise in temperature}}{\text{moles of metal}} \times 4.18.$$

The assumption was made that the thermal capacity per cm³ of the solutions involved did not differ greatly from unity. The results of some rough experiments on this appear to justify the assumption within the limits of experimental error.

To find the heat of reaction with acid, the experiment was repeated in the same way as before, using similar quantities of metal but adding them to 50 cm³ of standard acid. No cooling correction is necessary in this part. The final solution was made up to 500 cm³ and titrated as before.

To get reasonable results the experiment must be carefully performed. The maximum accuracy is about 5 per cent, being limited by the comparatively small rise in temperature on neutralization. However, the errors appear to be random and an average of a number of results gives good agreement with the law. A specimen result using calcium:

mass of calorimeter	= 46 g
volume of water	= 50 cm³
temperature of water before adding calcium	= 12.8 °C
temperature of water after adding calcium	= 24.8 °C
time of reaction	= 2½ min
temperature drop per min at 24.8 °C	= 0.5 °C
∴ average temperature drop per min	= 0.25 °C
∴ corrected final temperature	= 24.8 + 0.25 × 2.5
	= 25.4 °C
∴ temperature rise	= 12.6 °C

25 cm³ of 1.49 M acid added, both at 11.3 °C

$$\text{final temperature} = 14.8 \text{ °C}$$

$$\therefore \text{temperature rise} = 3.5 \text{ °C.}$$

It is found that 25 cm³ of this acid added to 50 cm³ of water in this calorimeter produces 0.3° rise

$$\text{corrected rise} = 3.2 \text{ °C.}$$

Contents of calorimeter made up to 250 cm³ and titrated against 0.099 M/2 Na_2CO_3. Solution found to be 0.0852 M

$$\therefore \text{moles of acid finally present} = 0.0213$$

$$\text{moles of acid originally present} = 0.0373$$

$$\therefore \text{moles of calcium added} = \frac{0.016}{2}$$

$$\therefore \text{heat of reaction with water} = \frac{54.6 \times 12.6 \times 10^{-3}}{0.016} \times 2 \times 4.18$$

$$= 359.5 \text{ kJ mol}^{-1}$$

$$\text{heat of neutralization of Ca(OH)}_2 = \frac{79.6 \times 3.2 \times 10^{-3}}{0.016} \times 2 \times 4.18$$

$$= 133.8 \text{ kJ mol}^{-1}.$$

Using the same calorimeter containing 50 cm³ of 1.49 M HCl

$$\text{original temperature} = 13.3 \text{ °C}$$

$$\text{final temperature} = 29.1 \text{ °C}$$

$$\text{temperature rise} = 15.8 \text{ °C.}$$

Contents made up to 500 cm³ and titrated as before.

$$\text{final molarity} = 0.12$$

$$\therefore \text{moles of acid finally present} = 0.06$$

$$\text{moles of acid added} = 0.0745$$

$$\therefore \text{moles of calcium} = \frac{0.0145}{2}$$

$$\text{heat of reaction} = \frac{54.6 \times 15.8 \times 10^{-3}}{0.0145} \times 2 \times 4.18$$

$$= 496.4 \text{ kJ mol}^{-1}.$$

Summary

(a) heat with water $= 359.5$ kJ mol⁻¹
(b) heat of neutralization $=$ 133.8 kJ mol⁻¹
(a) + (b) $= 493.3$ kJ mol⁻¹
(c) heat with acid $= 496.4$ kJ mol⁻¹.

In this experiment (a) was lower than average while (b) was higher, though both were within the usual range. This may be due to reaction (*a*) being incomplete.

The calorific value of coal gas

M. J. CLARK

With the increased emphasis now placed on energy as a fundamental physical concept, the need for a simple method of measuring the potential energy of a fuel becomes more apparent.

The method we have developed is not of course an original concept, but its use of inexpensive and readily obtainable materials permits the construction of sufficient units for a class experiment. Only a very elementary knowledge of calorimetry is required. Provided no gross leakage occurs there seems little danger associated with this experiment.

A known quantity of gas is stored in a polythene flask and then burnt with a Bunsen. The heat energy released is measured with a simple calorimeter.

The container is one of the common 'gallon' fruit juice flasks. It is calibrated for delivery using half-litre measuring cylinders full of water; the graduations are written with a 'Chinagraph' pencil. It is then *completely* filled with water and the filling-tubes inserted. The construction of the 'gasometer' and the filling and emptying sequence are illustrated in the accompanying diagrams. The syphoning action that occurs in *A* produces a reduced pressure in the container which increases the inflow of gas. The water level is allowed to fall well below the zero mark to allow for wastage as the Bunsen is lit. A clip is then tightened on to the rubber tubing between the adaptor and the container to prevent continuation of the syphoning action and mixing of the gas with air. A large calorimeter or empty 2 lb fruit tin is then filled with 250–300 cm^3 of water and placed on a tripod. The connections to the container are changed to configuration *B*, the clip removed, the water supply adjusted to give a moderate gas flow and the Bunsen lit. With arrangements as suggested a flame height of some 5 cm will be obtained. The Bunsen is placed under the calorimeter as the water level rises to the zero mark and is removed when the chosen amount of gas has been burnt. If the experimenter forgets to switch off the tap the Bunsen flame will be extinguished by a jet of water from within!

The water in the calorimeter is stirred while being heated and the initial and final temperatures measured. The calorific value in kilojoules per litre is found from the expression:

$$\text{Calorific value} = \frac{\text{mass of water} \times \text{rise in temperature}}{1000 \times \text{litres burnt}} \times 4.18.$$

In this approximate experiment the thermal capacity of the calorimeter can be neglected.

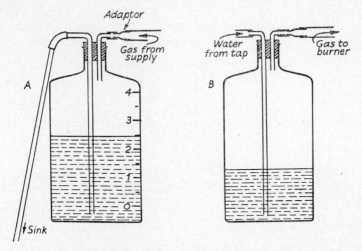

It is evident that a considerable proportion of the heat energy is lost to the atmosphere in this simple method. An intelligent class will tell you this but will appreciate the simplicity and directness and may be inspired to suggest modifications which would improve the accuracy of the experiment.

An experiment with 'meta' fuel
H. F. BOULIND

Tablets of 'meta' fuel may be purchased from chemists or camping stores; it serves two purposes; killing snails and boiling picnic kettles. It is also useful for laboratory experiments.

To illustrate the meaning of 'calorific value' of a fuel. In addition to the 'meta', each pair of pupils requires a tripod, a thermometer, and a small tin, and a lid—a golden syrup tin is excellent. 200 cm³ of water is measured into the tin, which is placed on the tripod. Half of a bar of 'meta' fuel is placed on the tin lid and weighed. A 100 g spring balance would do, but a beam balance is better. The lid is placed under the can of water, the temperature of the water is taken, and the 'meta' fuel is lit by a match. The fuel should be supported about 2.5 cm under the base of the can; it can be stood on another tin or tins. The temperature of the water is taken (°C) after the fuel has burnt out, then the lid and remaining 'meta' is weighed again. The calorific value in joules per g is,

$$\frac{\text{mass of water} \times {}^{\circ}\text{C rise of temp.}}{\text{burnt mass}} \times 4.18.$$

For comparison, the pupils could do a similar experiment with a candle, weighing the candle before and after heating the water, and thus comparing the calorific value of 'meta' fuel with that of wax. The class might then be asked if this is an accurate determination, and why not?—so leading, if the teacher thought fit, to a few words about a more accurate fuel calorimeter.

Experiment to measure the heat of combustion of sulphur
D. C. FIRTH

The sulphur can be firmly attached to the iron wire, by looping the wire and dipping it into molten sulphur, the wire being weighed before and after.

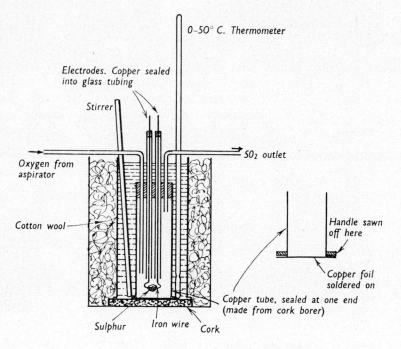

Care is necessary in connecting the wire to the electrodes since a little of the sulphur tends to break away. Oxygen is passed into the 'bomb' from an aspirator, and the outlet checked to make sure that the gas is circulating before the sulphur is ignited. A steady flow of gas is maintained throughout the combustion. Two 2 V accumulators can be used to start the burning. Apart from an increase in temperature, a successful ignition is shown by smoke issuing from the sulphur dioxide outlet.

Chemical interaction between the copper and sulphur dioxide must occur,

I suppose, but apart from pointing this out to the class I ignore it in the calculation. The escaping sulphur dioxide must carry away some heat, but it is small (circulating the gas through a copper spiral immersed in the water made little difference to the values obtained). The results shown have been obtained.

Mass of iron wire	$= 0.064$ g
Mass of wire $+$ S	$= 0.419$ g
Initial temperature	$= 16.8\ °C$
Final temperature	$= 18.1\ °C$
Mass of water	$= 560$ g
Heat capacity of apparatus	$= 40$ g
Mass S	$= 0.355$ g
Rise	$= 1.3\ °C$

Calculation:

$$\text{Heat gained by water} + \text{calorimeter} = (560 \times 1.3 + 40 \times 1.3) \times 4.18$$
$$= 3.26 \text{ kJ}$$

$$\therefore 0.355 \text{ g S produce } 3.26 \text{ kJ}$$
$$32 \text{ g S produce } 294 \text{ kJ}$$

Four determinations have shown values varying from 272 to 297 kJ mol^{-1}.

In the calculation no allowance is made for the oxidation of the iron wire because in practice the wire remains. In two determinations the loop remained intact, in two it was found to be broken.

The thermometer could be read to $0.05\ °C$. An error of $0.05\ °C$ in reading the temperature would give an error of ± 12 kJ mol^{-1} in the determination.

Heat of reaction

M. E. CHARLESWORTH and E. M. PATCH

The following can be carried out as a class experiment, the apparatus required being cheap and simple. The heat capacity of the Thermos flask was neglected.

Experiment

To determine the heat evolved when 1 mole of zinc displaces copper from a solution of copper(II) sulphate(VI).

Process

Weigh out about 4 g of finely granulated zinc in a test-tube which just fits into the neck of a cheap Thermos flask. Make up about 200 cm^3 of a solution of copper(II) sulphate(VI) containing between 30 and 40 g of the salt (this being

EC—C

enough to ensure excess for the reaction). Pour this, at room temperature, into the Thermos flask; suspend the test-tube in the liquid; insert an ordinary thermometer, plug up the neck with cotton-wool and leave for about half an hour to assume constant temperature. Record the temperature, pour the zinc into the solution and stir with the thermometer for about 10 minutes, to allow time for complete reaction, keeping the neck still plugged. Record the rise in temperature.

$$Zn(s) + Cu^{2+}(aq) \rightarrow Zn^{2+}(aq) + Cu(s); \ \Delta H = -209.5 \text{ kJ mol}^{-1}$$

The following results were actually obtained:

Mass of zinc	$= 4$ g
Volume of copper(II) sulphate(VI) solution	$= 200$ cm^3
Temperature before mixing	$= 19$ °C
Temperature after mixing	$= 34$ °C
Rise in temperature	$= 15$ °C
Reaction of 4 g zinc produced	3000×4.18 J
Reaction of 65.4 g zinc produced	$= 204.0$ kJ mole^{-1}
Value from tables	$= 209.$kJ mol^{-1}

Calorimetry of the hand

R. L. PAGE

One of the difficulties of teaching calorimetry is to make it relevant to the class's experience. The following experiment overcomes this difficulty in part, by measuring the heat lost by the hand. The hand is a reasonable source—on average it gives out about 8 kJ in ten to fifteen minutes. Placed in a 1000 cm^3 beaker it can be covered normally, by about 400 cm^3 of water, and thus over a period of ten to fifteen minutes a temperature rise of 4 to 5 °C can be obtained. A cut-away diagram of the apparatus is shown below in Fig. 1. The jacket is made of corrugated cardboard, and the base on which the beaker stands, of asbestos sheet or enamel tile. For slightly more accurate results a $\frac{1}{5}$ °C thermometer can be used, but for many groups an ordinary thermometer will be good enough.

A typical set of results gained from a group of PE students is as follows:

Mass of water to cover hand	$= 400$ g
Temperature before	$= 15.4$ °C
Temperature after	$= 20.4$ °C
Temperature rise	$= 5.0$ °C
Heat gained by water and given out by hand	$= 8.36$ kJ
Time taken	$= 15$ minutes
Heat given out/minute	$= 0.56$ kJ min^{-1}

To find the surface area of the hand to a fair degree of accuracy, take a sheet of fairly thick cardboard. Draw an outline of the open hand on it up to the water line, and cut it out. Then cut a square 10 cm × 10 cm. Weigh both of them, M_1 and M_2, respectively (Fig. 2).

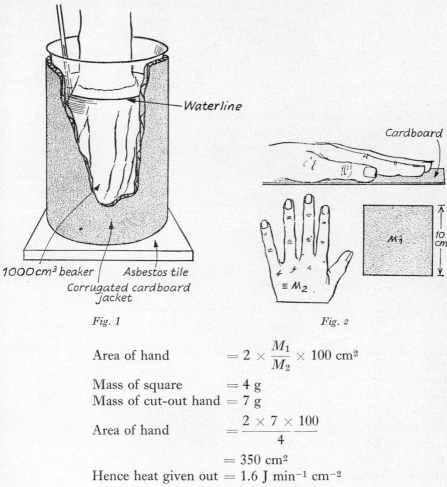

Fig. 1   Fig. 2

Area of hand $\quad = 2 \times \dfrac{M_1}{M_2} \times 100 \text{ cm}^2$

Mass of square $\quad = 4$ g

Mass of cut-out hand $= 7$ g

Area of hand $\quad = \dfrac{2 \times 7 \times 100}{4}$

$\qquad\qquad\qquad = 350 \text{ cm}^2$

Hence heat given out $= 1.6$ J min^{-1} cm^{-2}

I have found so far a figure ranging from 0.8 to 2.5 J min^{-1} cm^{-2}. After strenuous activity (a PE lesson or some other athletic activity) this value can rise by as much as 0.8 J min^{-1} cm^{-2}. This experiment is quite a useful introduction to heat loss from the body, which in turn can lead to heat production to keep the body temperature steady, dieting, etc.

It should be pointed out that this experiment measures heat loss in water which is more than heat loss in air. That is, the experiment approximates to heat loss while swimming rather than walking, etc. None the less I think it brings calorimetry a little nearer to a child's experience.

Measuring the energy of the carbon–bromine bond

IAN KNOX

One topic studied in the recent Department of Education and Science course for teachers of chemistry held at University College, London, was the concept of 'bond energy terms'; these may then be used to predict a heat of reaction, for example. Professor D. J. Millen suggested that the addition reaction between bromine and a liquid alkene such as cyclohexene might be suitable for use in schools to find the carbon–bromine bond energy term. The advantage of this system is that both of the reactants are readily available liquids.

We therefore investigated this reaction and found it to proceed smoothly and exothermically, and from the temperature rise we were able to deduce the molar heat of reaction. The method we used was as follows: a suitable

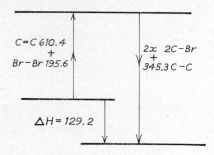

Fig. 1. Energy cycle for reaction (all values in kJ mol⁻¹)

volume of a solution containing about 10 g bromine in 1 dm³ of carbon tetrachloride was poured into a Dewar flask fitted with a stirrer, a heating coil, and a sensitive thermometer. Because of evaporation, the temperature of the solution fell slowly. Since a temperature–time curve was plotted, this cooling did not lead to error in obtaining the temperature rise. A small excess of cyclohexene, about 1–2 cm³ in our case, was added, and temperature readings continued until the new cooling curve could be plotted. The temperature rise in our case was about three degrees. A current was then passed through the heating coil so that a similar temperature rise was produced in the system in a similar time. From this, the heat produced by the reaction involving a known mass of bromine was calculated, and hence the molar enthalpy of reaction was found.

The bromine concentration was found by direct weighing, cooling the weighing bottle in dry ice to minimize loss of bromine. Alternatively, an iodimetric titration could have been used to estimate the bromine solution. The electrical energy used was obtained from the i^2Rt formula: some schools might use a joulemeter here. Alternatively, the enthalpy of reaction could be calculated from an estimate of the thermal capacity of the system, without using the electrical heater.

The experiment measures the enthalpy of the reaction:

cyclohexene + bromine = 1,2-dibromocyclohexane

and it can be seen that in total a carbon–carbon double bond is changed to a single bond, a bromine–bromine bond is ruptured, and two carbon–bromine bonds are made. The bond energy terms of C=C, C—C and Br—Br are well recognized and we took them as 610.4, 345.3 and 195.6 kJ mol^{-1} respectively. We found the enthalpy of reaction to be 129.2 kJ mol^{-1}.

If the C—Br bond energy term is called x we have:

$$610.4 + 195.6 + 129.2 = 2x + 345.3 \text{ (from Fig. 1)}$$
$$2x = 589.9$$
$$x = 295 \text{ kJ mol}^{-1}.$$

Activation energy

Experiment to show the principle of the Davy lamp

F. FAIRBROTHER

A copper gauze cylinder, closed at one end, is made from a piece of gauze about 12 mesh, 16.5 cm × 15.0 cm. This produces a cylinder 13.7 cm high and 3.8 cm diameter, with 1.9 cm overlap. Along the top edge V-cuts are made 1.3 cm down. To close the top, a disc of gauze is inserted and the top edges folded over. The cylinder is bound with a few turns of bare copper wire.

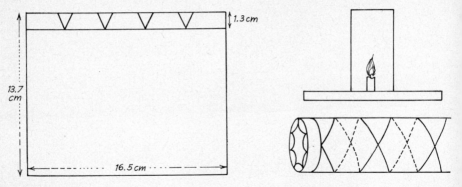

The cylinder is pressed over a piece of candle, 4 cm long, in a circle of Plasticine, 10 cm diameter and 1.3 cm thick. The gas from a Bunsen burner can be played on the outside of the cylinder while the candle is alight inside. The gas can be observed burning inside the gauze, but the burner is not ignited.

Repeat, after making a small depression or tunnel in the Plasticine with a glass rod; the Bunsen ignites.

The catalytic oxidation of alcohols using the principle of the safety lamp

GEORGE NOVELLO COPLEY

Teachers of organic chemistry are probably familiar with the simple form of 'formaldehyde lamp', in which a red-hot platinum spiral held near to the surface of some warm methanol in a beaker continues to glow owing to the exothermic nature of the oxidation of methanol to methanal and formic acid:

$$CH_3OH + air \rightarrow HCHO \rightarrow HCOOH.$$

I have found that the chief trouble in demonstrating this thaumaturgic instance of surface catalysis is in preventing the methanol vapour from catching fire. Accordingly, I have now devised an apparatus which is very easy to assemble, overcomes this difficulty by using the principle introduced by George Stephenson and Sir Humphrey Davy in their miners' safety lamp, and has certain other advantages as well.

The apparatus is shown in Figs. 1*a* and 1*b*. The alcohol to be oxidized is placed in the small porcelain crucible *a*, which is covered with a copper-wire

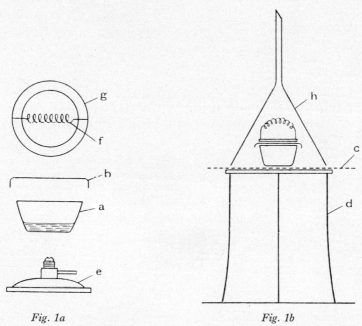

Fig. 1a   Fig. 1b

gauze *b* (12 meshes to the linear cm), and rests on the same kind of gauze *c*, resting in turn on a tripod *d*. The alcohol is heated by the lower part of a Bunsen *e*. A platinum spiral *f* supported by a lead ring or large washer *g* can be placed on the gauze *b* and the whole covered with a 10 cm glass funnel *h* in the manner shown in the figure *b*. The stem of the funnel is connected by rubber tubing to a drechsel bottle and an aspirator or water-pump which are not depicted in the figures. The water-pump is allowed to run at a rapid rate so that plenty of air is drawn through the inverted funnel. The alcohol is meanwhile warmed, and when it is judged to be giving off sufficient vapour the platinum spiral is heated in a Bunsen flame, placed on the gauze *b*, followed immediately by the funnel. The spiral *f* will now glow almost indefinitely; if it does not, then there is probably too much alcohol vapour in the funnel and the heating by means of e should be interrupted. Screening the apparatus from draughts is also an effective precaution to take.

The following alcohols, primary, secondary, and tertiary among them, have been found to undergo oxidation in this apparatus; those marked with an asterisk appear to work best; terpineol, a fairly involatile tertiary alcohol, works worst. Phenylmethanol (benzyl alcohol) and ethane-1,2-diol (glycol) had no effect at all. Methanol, ethanol, propan-1-ol, propan-2-ol,* butan-1-ol, 2-methylpropan-1-ol, 2-methylpropan-2-ol,* pentan-1-ol, cyclohexanol,* terpineol.

A little Schiff reagent placed in the drechsel bottle will indicate the formation of aldehydes in these oxidations. I am now endeavouring to extend the use of this kind of apparatus to the characterization of alcohols, by having a solution of 2,4-dinitrophenylhydrazine in the drechsel bottle and so obtaining a 2,4-dinitrophenylhydrazone of the aldehyde or ketone produced.

The experimental verification of the Arrhenius temperature–reaction rate relationship and the determination of an energy of activation

R. W. RICHARDSON

Since the time taken to complete a given reaction is inversely proportional to its rate constant it is sufficient to prove that a linear relationship exists between $\log t$ and $1/T$, where t is the time for the reaction to take place at a temperature T K.

The experiment described shows that this relationship is closely obeyed for the reaction

$$5Br^- + BrO_3^- + 6H^+ = 3Br_2 + 3H_2O$$

within a temperature range of 20–70 °C.

To an acidified bromide-bromate solution is added a small stoichiometrical deficiency of phenol and a few drops of methyl red solution. As soon as the bromination of the phenol has been completed the excess bromine decolorizes the indicator. If constant amounts of bromide-bromate, acid and phenol are mixed the complete formation of 2,4,6-tribromophenol represents the liberation of a given amount of bromine.

Experimental details. Reagents and apparatus required.

> Bromide-bromate(v) solution: $M/60$
> Sulphuric(vi) acid: 0.25 M
> Phenol solution: 1.1–1.2 g dm^{-3}
> Methyl red: 1 g dm^{-3} alcoholic BDH indicator.

A thermostat or any large reservoir capable of maintaining a temperature of ± 0.5 °C.

Method

The following procedure is carried out at intervals of $10°$ from $20-70$ °C. 10 cm^3 each of the bromide-bromate(v) and phenol and 3–4 drops of methyl red solution in one flask and 50 cm^3 of the dilute acid in another are brought to the required temperature in the thermostat. The acid is then quickly poured into the bromide-bromate(v) solution and the time taken from that instant to the complete decolorization of the indicator found.

The logarithm of the time taken for the decolorization of the methyl red is then plotted against the reciprocal of the absolute temperature of the experiment.

The following results were obtained on solutions which on analysis gave

Bromide-bromate(v): 0.01636 M
Sulphuric(vi) acid: 0.25 M
Phenol: 1.19 g dm^{-3}

°C	T/K	1/T	t/s	$\log_{10} t$
20	293	3.413×10^{-3}	296	2.471
30	303	3.301 ,,	133	2.124
40	313	3.196 ,,	67	1.827
50	323	3.096 ,,	32	1.504
60	333	3.003 ,,	15	1.176
70	343	2.916 ,,	8	0.903

The accompanying graph shows the linear relationship between $\log_{10} t$ and $1/T$.

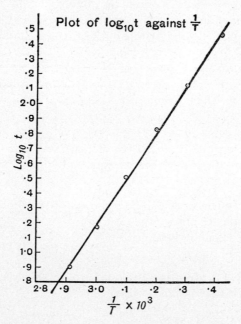

Plot of $\log_{10}t$ against $\frac{1}{T}$

The calculation of the energy of activation (E)

Since the rate constant (k) is inversely proportional to t the Arrhenius equation,

$$\log_e k = -\frac{E}{RT} + C$$

may be written

$$\log_{10} t = \frac{E}{2.303RT} + C',$$

then the slope of the graph (m)

$$= \frac{E}{2.303R}$$

and

$$E = 2.303Rm.$$

The figures above give E as 58.3 kJ for the reaction considered.

The activation energy of the thiosulphate–acid reaction

J. W. TURNER

The use of the reaction between sodium thiosulphate and dilute hydrochloric acid in the teaching of elementary reaction kinetics is well established [1, 2, 3, 4]. An inverse measure of reaction rate is obtained by measuring the time taken for a given amount of reaction to take place, i.e. the time for a given amount of colloidal sulphur to be precipitated, so that it just obscures an ink spot placed behind the reaction vessel. Using this technique, the variation rate with (a) thiosulphate concentration, and (b) temperature, may be investigated. The results obtained in (b) above produce a satisfactory Arrhenius plot, from which the activation energy is evaluated.

THEORY

The rate of a reaction as measured by the rate constant k is related to the absolute temperature by the Arrhenius equation:

$$k = A.e^{-E/RT}$$

where A = the frequency factor, E = the activation energy, R = the gas content and T = the absolute temperature.

Taking logs, the equation becomes

$$\log_{10} k = \log_{10} A - E/2.303RT.$$

The time t taken for a given quantity of sulphur to be formed is inversely proportional to the rate constant k, thus

$$\log_{10} t = E/2.303RT + \text{constant}$$

hence a plot of $\log_{10} t$ against $1/T$ will give a straight-line graph whose gradient is $E/2.303R$.

PROCEDURE

Attach a piece of paper marked with a dark ink spot to the side of a litre beaker. Fill the beaker with water and clamp two boiling tubes A and B vertically in the beaker so that they are about half immersed. Position both the beaker and tube A so that the ink spot is on the opposite side of the beaker to the observer and can be seen through both the beaker and the tube A. Add 10 cm³ of 0.1 M thiosulphate to tube A and 10 cm³ of approximately 0.5 M hydrochloric acid to tube B. Allow both to attain thermal equilibrium with the surrounding water and then quickly add the acid from B to the thiosulphate in A. Stir well with a thermometer and record the steady temperature. Using a stopwatch, note the time taken for the dark spot to just become completely obscured by the formation of colloidal sulphur. Repeat the experiment at several different temperatures, say at 10, 15, 20, 25, 30, 35, 40, 50 and 60 °C, using the same strength solutions and observation technique.

RESULTS

The following results, summarized graphically, in Fig. 1 were obtained by a student during a single practical session and yield an activation energy of 46.58 kJ for the reaction.

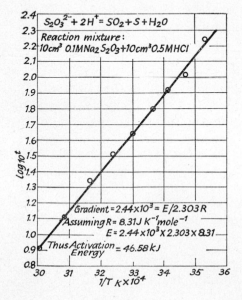

Fig. 1

REFERENCES

1. Nuffield Foundation, *Chemistry Sample Scheme, Stages I* and *II* (Longman/Penguin, 1966), Sections 18.2 and 18.3.
2. Abbott, D., *An Introduction to Reaction Kinetics* (Longmans, 1966), p. 3.
3. Atherton, M. A. and Lawrence, J. K., *An Experimental Introduction to Reaction Kinetics* (Longmans, 1940), pp. 16 and 157.
4. Cleeve, H. N., *S.S.R.*, 1970, 179, **52**, 370.

Activation energy determinations

M. A. MACKINNON

When a solution of potassium chromate(VI) is added to a solution of a simple salt of either iron(III), cobalt(II), nickel(II) or copper(II), a brownish precipitate of a basic chromate is formed more or less slowly. Explanations of the reaction in the literature are best described as confusing.

With copper(II) and iron(III) salts, precipitation is rapid at room temperatures unless solutions are very dilute. ($FeCl_3$ is too acid to precipitate at all, and $Fe(II)$ is of course oxidized.) However, with cobalt and nickel the formation is conveniently slow for a very simple series of experiments to measure the activation energy of the reaction. It is nearly the same in both cases (class experiments here have given average values of 109.9 and 112.4 kJ respectively), but the nickel reaction is much slower. This allows profitable discussion of the effects of the parameters A and E in the Arrhenius expression: reaction rate $= A.e^{-E/RT}$, a discussion which otherwise is difficult to relate to solutions in flasks.

Practical details

We have used solutions approximately 5 per cent, that is 50 g dm^{-3} potassium chromate(VI) and 70 g dm^{-3} either cobalt(II) nitrate(V) or nickel(II) sulphate(VI). Solutions are put into burettes. 20 cm^3 chromate(VI) is run into a conical flask standing on newspaper, and the same volume of cobalt(II) salt is run into a polythene beaker. The time is noted as the solutions are mixed and swirled, and again when the print becomes quite illegible. The reaction has a small enthalpy change, but the temperature must be measured fairly exactly because the activation energy is so large. Typical approximate times for illegibility are about 2 minutes at 7.5 °C down to 10 seconds at 22 °C. We have then tabulated the reciprocal of the illegibility time in seconds^{-1} and plotted the ln of this figure against the reciprocal of the absolute temperature. E can then be calculated from the slope.

With nickel(II) sulphate(VI) the reaction is very slow at room temperature and this means that both solutions, and the flask, must be warmed before mixing.

Practice and several repeat performances are needed to obtain reliable results, but the experiment is simple and rapid. Typical times for illegibility of the print are 70 seconds at 43 °C and 25 seconds at 50 °C.

The results with iron(III) and copper(II) salts are rather unsatisfactory since the dilution has to be so large to bring the rate low enough for easy measurement. The precipitate is then less powdery and the disappearance of the print difficult to judge. It is tempting also to vary the concentrations of the cobalt(II) and chromate(VI) solutions, keeping the total volume to 40 cm³, in order to study the order of reaction, but this is rather unconvincing too; if the concentrations are similar the reaction appears first order for cobalt(II) ion and roughly first order for chromate(VI) (better agreement if order is about 1.3). With other concentrations the results are fairly incomprehensible; it is clear that the product is not one simple compound.

Note. The precipitate sticks to glass if it is allowed to form in quantity, but a trace of HCl removes it instantly.

Experiments for the sixth form student; aspects of energy changes

A financial analogue of the laws of thermodynamics

H. L. ARMSTRONG

The first and second laws of thermodynamics may be expressed in the respective equations

$$dU = dq - dw$$
$$dS = dq/T$$

along with the remark that dU and dS are exact differentials. The symbols U and S, of course, represent respectively internal energy and entropy, while q represents heat absorbed by the system, w work done by the system, and T the temperature. Instead of using just this form, one may make statements, which amount to the same thing, about the vanishing of integrals around closed cycles.[1]

The quantities U and S are rather abstract things; a concrete analogue to the situation may therefore be helpful. The following one, involving money —a subject of well-nigh universal interest—is suggested.

Consider, then, a company (say), which buys supplies and sells products, and makes and receives payments. One could say that, during any interval

(change in total assets) = (cash received)
    — (value of goods shipped).

In this reckoning, cash disbursed will count as cash received, but with a negative sign, and the value of goods or supplies purchased will count as 'value of goods shipped', but again will be counted as negative.

It is more difficult to give an analogue for the second law, but something may be done in the following way. Suppose that the company being considered makes and receives all its payments in cash, there being no cheques at all; and, moreover, that each payment is made in just *one* piece of money, i.e. *one* note or *one* coin. Then one could say

(change in number of pieces of money in the till)
    = (cash received)/(denomination in which the payment was made)

Why anyone should want to keep track of the number of pieces of money

[1] H. L. Armstrong, *Am. J. Phys.*, **28**, 564 (1960).

in the till is not discussed, although doubtless the development of an expensive electronic computer, to compile that information and store it on punched cards, would be hailed as a great advancement in something or other.

These analogues illustrate something which is not always very clear: that, as far as thermodynamics goes, the internal energy and the entropy are just *book-keeping* functions, useful in keeping track of the changes in things directly apparent to us. Just as one would not expect to go into a vault and see the total assets of a company piled up there, so he should not think of the internal energy as 'something in' a system, but as merely an aid in compiling 'something that has been done' to the system. Similar remarks, of course, apply to the entropy; to give either of these quantities significance as other than book-keeping functions, one must leave the boundaries of strict thermodynamics and call on kinetic theory for help.

A heat engine run by rubber
PROFESSOR E. G. COX

For a fixed extension not exceeding about 350 per cent the tension in a strip of rubber is approximately proportional to the absolute temperature over the range 200–400 K. This result, with its obvious analogy to the simple gas law, provides direct experimental support for the kinetic statistical theory of rubber elasticity, for it shows[1] that, approximately, over the range stated, the internal energy of rubber at a constant temperature is independent of its extension and that the increase in tension with stretching is due almost entirely to decrease of entropy. The 'perfect rubber' is therefore an interesting alternative to the perfect gas as a subject for exercises in elementary thermodynamics; as for a gas, a simple kinetic explanation of the facts can be given, but rubber has the advantage of being more easily used as the working substance in a simple heat engine suitable for lecture demonstrations. Such an engine was apparently first described by Wiegand, who constructed a pendulum and a rotary 'rubber motor',[2] in both of which continuous motion was achieved by the alternate heating and cooling of stretched rubber. It is however, not an easy matter to obtain consistently good results with these devices, and I have found that the considerably simpler arrangement described below, which is similar to the muscle motor of Prior,[3] is much more satisfactory.

The motor consists of a wheel with a ball-bearing hub, rubber spokes, and a rigid light metal rim 20 cm in diameter, arranged to rotate about a fixed

[1] See, e.g., Treloar, L. R. G., *The Physics of Rubber Elasticity*, Oxford University Press, London, 1949, Chapter II.
[2] Wiegand, W. B., *Trans. Inst. Rubber Ind.*, **1**, 164–5 (1925).
[3] Prior, M. G. M., *Nature*, **171**, 213 (1953).

horizontal axle. If radiant heat is applied to the lower spokes of the stationary wheel they contract slightly, thereby raising the centre of gravity of the rim above the fixed axle, so that the wheel turns, bringing fresh spokes into the radiation and allowing those previously heated to cool. By suitable adjustment of the heated area continuous rotation in one direction can be maintained. About 60 ordinary elastic bands at 300–400 per cent extension are used as spokes; a rather large extension is necessary to prevent sagging of the rim. In order to equalize tensions in opposite spokes each band was stretched across a diameter and the hub was clamped to the spokes at the centre of the wheel by a washer and nut. Final balancing of the wheel was achieved by adjusting small pieces of Plasticine on the rim.

Heat is applied by placing a 60-watt lamp bulb close to the lower spokes and somewhat to one side of the vertical through the axle; a reflector is placed opposite the lamp behind the wheel so as to increase the heating effect by throwing back some of the radiation which has passed through the spokes. The wheel then turns at a rate which can be regulated to some extent by 'advancing' or 'retarding' the heater but which is of the order of magnitude of one revolution per second; we have not had occasion to run it continuously for more than two hours but there seems no reason why it should not run for much longer periods. It ran successfully as first made with materials which happened to be available so that there is apparently nothing very critical about the details of design or construction. An improvement suggested to me by Mr D. W. Saunders is to coat the surfaces of the rubber spokes with carbon black to increase the absorption and radiation of heat.

We have also been able to produce rotation by the application of cold to upper spokes instead of heat to the lower ones; with the arrangement used (a stream of air of a few square centimetres cross-section, cooled by passing through carbon dioxide snow) the motion was slow and irregular, but with attention to details of the cooling arrangements better results could no doubt be achieved.

I am indebted to Mr V. Balashov for assistance in the construction of this apparatus.

The thermodynamics of rubber
PROFESSOR E. G. COX

If we allow a filled rubber balloon to collapse suddenly we can feel that the issuing air is cooled appreciably: if we could make suitable arrangements for the observation we should find that the rubber also is cooled. The amounts of heat involved are comparable: if the air has been compressed to about 2 atmospheres and the rubber extended about 500 per cent the heat absorbed as they return to normal is of the order of 40 J g^{-1} in each case. These and other similarities between the expansion of a gas and the contraction of rubber (or any rubber-like polymer) are due to the fact that many of the properties of both stem from the kinetic agitation of their molecules, and the 'perfect rubber' is therefore an interesting variant to the 'perfect gas' as a subject for elementary exercises in thermodynamics. The replacement of pressure by tension, with the accompanying change of sign, is sufficient to make mere substitution in the gas equations of little avail, and to make the student think carefully when he is asked to derive the various results. Although the concept of entropy may be beyond the grasp of the average sixth-former, the study of this topic should at least enable him to see that the 'elasticity' of rubber is entirely different in its nature from the true elasticity of normal materials such as, e.g., steel or silk in which the main change accompanying stretching is in the energy of the material and is due to work being done against intermolecular forces as the relative positions of the atoms are changed.

The heat motions in rubber consist mostly of lateral vibrations of the atoms and rotations of groups of atoms in the long chain molecules of which the rubber is composed. The motions give rise to a 'repulsion pressure' which expands the material in directions at right-angles to the lengths of the molecules and which is therefore equivalent to an internal tension parallel to their lengths. The internal tension persists until the molecules are kinked up in random fashion with molecular impacts distributed equally in all directions. By the same train of reasoning as we use for the perfect gas we can show that in a perfect rubber the tension, at fixed extension, is proportional to the absolute temperature and that the internal energy is independent of the extension, so that all the work done in stretching the rubber is converted to heat. (It is necessary to assume that rubber undergoes no change of volume on stretching; this is very nearly true, Poisson's ratio being about 0.49 instead of $\frac{1}{2}$ K.)

It was shown experimentally by Meyer and Ferri that up to about 400 per cent extension, and over the range (roughly) 200–400 K, the tension in a piece of rubber at fixed extension is in fact approximately proportional to the absolute temperature. It is not at all difficult to design an experiment to verify this in an elementary teaching laboratory. It is also quite easy to make a 'rubber engine' to demonstrate the thermodynamic properties of rubber qualitatively; various types have been described but perhaps the simplest is the one I once displayed at the Annual Meeting in Leeds. This consists of a wheel with rubber spokes, exposed over part of its lower portion to radiation from an electric light bulb; the lower spokes contract as they are heated, so that the centre of gravity of the rigid rim is lifted above the axle, and the wheel rotates. With very little preliminary adjustment such a wheel can be made to rotate continuously for many hours and probably many weeks.

REFERENCES

Gee, G., *Quarterly Reviews of the Chemical Society*, **1**, 273, 1947.
Treloar, L. R. G., *The Physics of Rubber Elasticity*, Oxford University Press, 1949.
Meyer, K. H. and Ferri, C., *Helvetica Chimica Acta*, **18**, 570, 1935.
Cox, E. G., *Journal of Chemical Education*, 307, 1954.

ΔG or $\Delta G^{\ominus}$? Introducing standard free energy in the energetics course

N. J. SELLEY

It has become fashionable to include the concept of free energy in school chemistry courses, in some cases at O-level and almost certainly in the sixth form. This trend has been accompanied by many valuable attempts to simplify and adapt a traditionally difficult and 'mathematical' topic, without distorting and falsifying it.

In order to make thermodynamics comprehensible in school (and college, even) it has been necessary to play down the logical development of the equations, and to emphasize the application of these equations to real chemical situations. The resourcefulness of teachers has been exercised to find simple but meaningful experiments suitable for class use, to illustrate the principles and make the subject come alive. The changes in approach have been profound enough to warrant a new name, 'energetics'.

In some cases, however, the simplification seems to have gone too far, to the point of creating a nonsense. The failure to distinguish between standard and non-standard free energy is perhaps an example of this.

It may well be that first introduction of free energy to pupils is difficult enough already, without adding the complication of standard concentrations. Therefore it is probably justifiable to keep to ΔG at first, and to discuss the equation

$$\Delta G = \Delta H - T\,\Delta S$$

without emphasizing that ΔG and ΔS vary with concentration. This enables the class to go ahead with an investigation into the driving force of spontaneous changes, the reason why endothermic reactions can occur, and the reason why thermal decompositions (ΔH and ΔS both positive) become more feasible at higher temperature.

The danger is that the teacher may be tempted to follow up this success, and to press on to quantitative treatment, and even equilibrium constants, without revealing that ΔG is a function of concentration. Ultimately the pupil may be faced with the task of reconciling logical contradictions such as the following:

1. $\Delta G = 0$ at equilibrium

2. $\Delta G = -RT \ln K$
3. the equilibrium constant need not equal one.
   or
4. reaction is spontaneous if $\Delta G < 0$
5. the reaction $NH_3(aq) + H_2O \rightarrow NH_4^+ + OH^-$ occurs even though ΔG (meaning $\Delta G^{\ominus}$) is positive.

Clearly the idea that free energy varies with concentration of reactants *and products* (very important, this) cannot be postponed for long. After that the need for a *standard* free energy becomes apparent, and $\Delta G^{\ominus}$ presents no further difficulty.

The pupil can be led to discover these facts by a simple experiment—the concentration cell—which will be described shortly.

Before that, though, let us consider the experiments which are often performed at an early stage in the course. Even without an explicit distinction between ΔG and $\Delta G^{\ominus}$ it is possible to show that the work of reaction is not necessarily equal to the heat of reaction. ΔG can be measured electrochemically for a reaction occurring in a cell, with specified reactant and product concentrations.[1] ΔH for the same reaction can be measured calorimetrically.

The reaction $Zn(s) + Cu^{2+}(aq) \rightarrow Cu(s) + Zn^{2+}(aq)$ has been suggested for this purpose, but in the author's opinion it is unsuitable, since ΔH and $\Delta G^{\ominus}$ are closer in value than the experimental error. The reaction between $Cu(s)$ and $2Ag^+(aq)$ is preferable, despite the high cost of silver nitrate(v), because as $\Delta G^{\ominus}$ and ΔH are about 87 and 147 kJ g-eq^{-1} respectively, the difference is observable even if the error in each determination is as great as ± 10 per cent.

Regarding the cost, there is really no need to use the extravagantly concentrated 1 M silver nitrate(v), especially for the cell. If the silver nitrate(v) and copper(II) sulphate(vI) solutions are both diluted to 0.1 M the cell potential is only decreased by 0.03 V, or 6 per cent, which is probably no more than the experimental error if a voltmeter, rather than a potentiometer, is used. In the calorimetry part of the experiment, in which excess copper powder (reduced CuO) is reacted with silver nitrate(v) solution, the solution must be concentrated enough to give a measurable temperature rise: 0.3 M $AgNO_3$ gives a 5 K rise.

I would suggest that this comparison of ΔG and ΔH of reaction should be followed immediately by a very simple investigation of a concentration cell, to show the existence of free energy of dilution, and the need for specified (e.g. standard) concentrations. As this experiment is not yet familiar at introductory level, some details may be found useful.

---

[1] In practice the chosen concentrations are often 1 mole litre^{-1}, so that the pupils do in fact measure $\Delta G^{\ominus}$. ΔH is independent of concentration unless this is much greater than standard, and therefore $\Delta H = \Delta H^{\ominus}$.

Set up the cell in the diagram, initially with 1 M $CuSO_4$ solution in both half-cells. Keep the salt bridge out of the $CuSO_4$ solution except when readings are being taken. The wire is attached to each copper foil electrode by being passed through a hole near to one end, then twisted and squeezed by pliers. The identical electrodes are then connected to a millivoltmeter (or microammeter).

When the circuit is completed it is expected that, since the two half-cells are identical, there should be no p.d. between them. In practice a small deflection of the meter may occur, if it is very sensitive, and this should be recorded and regarded as the 'zero error' of the circuit. It is probably due to stresses in the metal electrodes, or liquid junction potentials, and should not be discussed with the pupils at this stage.

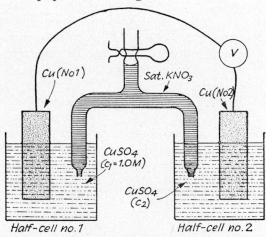

Fig. 1. *The simple concentration cell (from the author's* Chemical Energetics *by permission of Edward Arnold Ltd)*

The electrolyte in the half-cell number 2 is changed, first to 0.1 M $CuSO_4$, then to 10^{-3} M, and the maximum meter reading is noted each time. The polarity of the cell is observed. As this is crucial to the experiment, possibility of a mistake must be eliminated by tracing the wires from the electrodes to the terminals of the meter, and noting the polarity marked there by $+$ and $-$ signs, or by colours.

The meter readings are not cell potentials—a potentiometer circuit and a lot more trouble would be required for those—but they are proportional to them. The actual observations show that (1) the dilute half-cell is the negative electrode, and (2) the meter readings for 1 M/10^{-3} M is about three times, and not 100 times, the reading for 1 M/0.1 M.

In the discussion it will be deduced that the electrode reactions are:

positive (concentrated): Cu^{2+}(1 M) $+ 2e^- \rightarrow Cu(s)$
negative (dilute): $Cu(s)$ $\rightarrow Cu^{2+}$(dilute) $+ 2e^-$.

The overall reaction is just: Cu^{2+}(1 M) $\rightarrow Cu^{2+}$(dilute).

The experiment has proved that this dilution process is a spontaneous change, with a negative ΔG. The actual value of ΔG appears likely to be a function of $\log (c_1/c_2)$.

Further discussion might involve the connection between free energy of dilution and diffusion or osmosis, which are spontaneous dilutions. A more advanced class might see the connection with vapour pressure: for example they might consider a system consisting of two beakers, one of 1 M $CuSO_4$ and the other of 0.1 M $CuSO_4$, in a sealed box. The solution with the higher vapour pressure would evaporate, and the other would gain water from the air. The net result would be a concentration equalization similar to that occurring in the concentration cell.

In another direction, the discussion could treat the free energy of reactions which take place with non-standard concentrations. For example, the reaction:

$$Cu(s) + 2Ag^+(1\ M) \rightarrow 2Ag(s) + Cu^{2+}(10^{-3}\ M)$$

could be regarded as the sum of two steps:

$$Cu(s) + 2Ag^+(1\ M) \rightarrow 2Ag(s) + Cu^{2+}(1\ M) \quad \Delta G = \Delta G^{\ominus}$$
$$Cu^{2+}(1\ M) \rightarrow Cu^{2+}(10^{-3}\ M) \quad \Delta G = -\text{ve}$$

So the free energy of reaction becomes more negative if product concentrations are lowered. However, a comparison between the standard Cu/Ag cell potential (found in the earlier experiment) with the roughly determined concentration cell potential shows that the difference between ΔG and $\Delta G^{\ominus}$ is probably small for moderate deviations from standard concentration.

This is not the last time the pupils will meet the concentration cell. A year or two later they could profitably set up a potentiometer circuit and measure the cell potentials accurately, and show that they conform to the equations.

$$\varepsilon = \frac{-RT}{nF} \ln \frac{c_2}{c_1} \quad \text{and} \quad \Delta G = +RT \ln \frac{c_2}{c_1} \quad (c_1 = \text{'reactant' concentration}).$$

From this it is a short step to the Nernst equation and the van't Hoff isotherm. No new principles will be needed.

Practical measurement of entropy changes

C. J. MORTIMER

Some mention of entropy belongs in a modern A-level course, perhaps along the lines of J. A. Campbell's 'Why do chemical reactions occur?' (*New thinking in school science*, OEDC). However, the usual attempt to measure an entropy change in practice by comparing the enthalpy of the reaction.

$$Zn + Cu^{2+}(aq) = Zn^{2+}(aq) + Cu$$

determined calorimetrically with the free energy of the reaction found from the e.m.f. of $(Cu/Cu^{2+})/(Zn^{2+}/Zn)$ entails a small difference in two large quantities. In any case one would not expect a large entropy change in this reaction. The result is totally unconvincing.

I suggest the following as an alternative. If solutions 0.1 M Fe^{2+} and 0.1 M Fe^{3+} both made up from the ammonium sulphate double salts are mixed, any pupil will tell you that there is no reaction possible. They just mix. Now place the two solutions in two beakers connected by a salt bridge of, say, concentrated potassium chloride solution. Place a platinum electrode in each beaker and connect these to a high impedance voltmeter. There is an e.m.f. of about 0.35 V. If there is an e.m.f., energy can be obtained from the system. The direction of the e.m.f. makes it clear that the changes occurring are

$$Fe^{2+} \rightarrow Fe^{3+} + e^- \quad \text{and} \quad e^- + Fe^{3+} \rightarrow Fe^{2+}.$$

In fact, if the cell were allowed to run down, the solutions on each side would become the same. In practice this would take too long, so the e.m.f. is now measured between solutions containing, say, 90 cm³ Fe^{2+}(aq), 10 cm³ Fe^{2+}(aq) and 10 cm³ Fe^{2+}(aq), 90 cm³ Fe^{3+}(aq). A series of these ending with two solutions both 50 cm³ Fe^{2+}(aq) and 50 cm³ Fe^{3+}(aq) represent stages in the running down of the cell to zero e.m.f. in the last case. A graph can now be plotted of e.m.f. against percentage conversion of Fe^{2+} to Fe^{3+}. Typical results are shown below.

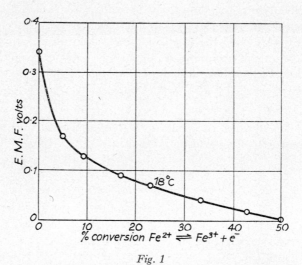

Fig. 1

Now conversion of 1 mole Fe^{2+}(aq) and 1 mole Fe^{3+}(aq) into solutions containing 50 per cent each in such a cell would require $\frac{1}{2}F$ coulombs. So the area under the graph represents joules of energy obtained as the cell runs down. From the results the energy obtainable from mixing 1 mole of Fe^{2+}(aq) with 1 mole Fe^{3+}(aq) is 3850 J at 18 °C.

If we assume that all the activities are the same then the entropy change expected

$$\Delta S = 2Nk \log_e 2,$$
$$= 2Lk \log_e 2,$$

(where L = Avogadro's constant, and k = the Boltzmann constant) since $2L$ particles have doubled their disorder, after mixing any particle selected has an equal chance of being Fe^{2+} or Fe^{3+}. Now

$$k = \frac{R}{L},$$

so
$$\Delta S = 2R \log_e 2$$
$$= 11.5 \text{ J deg}^{-1}$$

If $\Delta H = 0$ then
$$\Delta G = -\Delta S \times T$$
$$= -11.5 \times 291$$
$$= -3347 \text{ J}.$$

This is within 13 per cent of the experimental result.

Attempts to repeat the experiment at higher temperatures have resulted in excessive hydrolysis of the $Fe^{3+}(aq)$ ion with precipitation of iron(III) hydroxide, but higher e.m.f.s are found qualitatively as expected. This hydrolysis forming species such as $Fe^{2+}(OH)5H_2O$ may account for some of the 10 per cent error. One other obvious source of error in the simple treatment suggested is that on mixing the two solutions of 0.1 M Fe^{2+} and 0.1 M Fe^{3+} there are of necessity changes in the concentrations of anions, and at a concentration of 0.1 M this will presumably mean one cannot assume that the activity coefficients of the Fe^{2+} and Fe^{3+} ions remain constant.

It may be that other redox pairs would give 'better' results in a similar experiment.

Index